UNKNOWN FACTS OF OUR UNIVERSE

7 PARALLEL UNIVERSE

RUDRAKSH MISHRA

UNKNOWN FACTS OF OUR KNOWN UNIVERSE: THE 7 PARALLEL UNIVERSE

OUT OF THE BOX = Recently we found another planet which is the 11th planent......The name is VESTA...

As we all know that our universe has many galaxies and many galaxies has many planets.......We live on Earth which is a planet in MILKY WAY galaxy......We now know that there are 11 planets in our solar system.....But we dont the planets exept two geographical....We had not gone up there..."WE means scientists"....So that means we are unkown to our very known universe....May be there would be many planets like our Earth which is sufficient to sustain life.....

So parallel universe means a parallel dimension which ahs its co-existing with one's own....Any person has its own seven parallel universe...

LIVING IN A PARALLEL UNIVERSE= It is said that a person is living in parallel universe when his or her thoughts are different from allmost all the humans....The person who is said to live in a parallel universe thinks like a pro, their advises are almost accurate in the future...So they are said to live in future as they live in a parallel universe....

Actually there is also a another name of our universe that is bubble universe which represents an extra dimension.... In 2018 researchers have planned a new model of our universe which can solve enigma of a dark energy... { enigma=any thing or formula that is difficut to understand}....Our universe has 1 to 7 bubble universe.....Bubble universe got its name bubble because it reflects the features of our universe in it, means in other dimensions...

Arguing for a multiverse---

Around 13.7 billion years ago,, everything we were knowing in the cosmos was infinte singularity...Then according to the BIG BANG THEORY there must be an unkown souce which caused it to expand in the three dimensional space...

Eventually the small particles began to combine and form into larger pieces of matter that we know today such as stars etc... One big question that came in my mind after reading this theory was are we the only universe out there??

With our current technology we are not able to see such long distances.... There are at least five theories that i have read and the things that were common in all were - 1......We dont know the exact shape of space, maybe the universe is flat and endless as we know that it has infinte space..... 2....Eternal inflation { an outgrowth that is not understandable} based on a research of cosmos by cosmologist *ALEXANDER VILENKIN* that is also related with theory of BBT and at last my favourite and a popular physicist *STEPHEN HAWKING* has also dealt with multiverse in his last paper which was published in May 2018 he told that " WE ARE NOT DOWN TO A SIngle UNIVERSE BUT OUR FINDINGS IMPLY A REDUCTION OF MULTIVERSE TO MUCH SMALLER CHANCE OF ANY OTHER UNIVERSE".....

Arguing against a parallel universe----------------------

Not everyone agrees with parallel universe....The main problem for scientists to clearify this theory is that we dont know the age of universe....

You can see parallel universe in science fiction such as = MARVEL COMICE AND DC COMICS

Games such as Dungeons and dragons etc...

Let's understand multiverse a bit more...
The multiverse is a hypothetical group of multiple universes. Together, these universes comprise everything that exists: the entirety of space, time, matter, energy, information, and the physical laws and constants that describe them. The different universes within the multiverse are called "parallel universes," "other universes," "alternate universes," or "many worlds.

Early recorded examples of the idea of infinite worlds existed in the philosophy of Ancient Greek Atomism, which proposed that infinite parallel worlds arose from the collision of atoms. In the third century B.C. philosopher Chrysippus suggested that the world eternally expired and regenerated, effectively suggesting the existence of multiple universes across time. The concept of multiple universes became more defined in the middle ages.

In Dublin in 1952, Erwin Schrödinger gave a lecture in which he jocularly warned his audience that what he was about to say might "seem lunatic". He said that when his equations seemed to describe several different histories, these were "not alternatives, but all really happen simultaneously". This sort of duality is called "superposition."

The American philosopher and psychologist William James used the term "multiverse" in 1895, but in a different context..... The term was first used in fiction and in its current physics context by Michael Moorcock in his 1963 SF Adventures novella The Sundered Worlds (part of his Eternal Champion series).
Multiverse universe has been hypothesized in articles or books of cosmology , physics , quantum physics and mainly in science fictions... In all parallel universe is also called as alternate universe....The physics community has debated the various multiverse theories over time...

Prominent physicists are divided about whether any other universes exist outside of our own...

Some physicists say the multiverse is not a legitimate topic of scientific inquiry... Concerns have been raised about whether attempts to exempt the multiverse from experimental verification could erode public confidence in science and ultimately damage the study of fundamental physics...Some have argued that the multiverse is a philosophical notion rather than a scientific hypothesis because it cannot be empirically denied..... The ability to disprove a theory by means of scientific experiment has always been part of the accepted scientific method..... Paul Steinhardt has famously argued that no experiment can rule out a theory if the theory provides for all possible outcomes.....In 2007, Nobel laureate Steven Weinberg suggested that if the multiverse existed, "the hope of finding a rational explanation for the precise values of quark masses and other constants of the standard model that we observe in our Big Bang is doomed, for their values would be an accident of the particular part of the multiverse in which we live.......

SEARCH FOR EVIDENCE-

Around 2010, scientists such as Stephen M. Feeney analyzed Wilkinson Microwave Anisotropy Probe (WMAP) data and claimed to find evidence suggesting that our universe collided with other (parallel) universes in the distant past.....However, a more thorough analysis of data from the WMAP and from the Planck satellite, which has a resolution three times higher than WMAP, did not reveal any statistically significant evidence of such a bubble universe collision.... In addition, there was no evidence of any gravitational pull of other universes on ours....

In his 2003 New York Times opinion piece, "A Brief History of the Multiverse", author and cosmologist Paul Davies offered a variety of arguments that multiverse theories are non-scientific..

For a start, how is the existence of the other universes to be tested? To be sure, all cosmologists accept that there are some regions of the universe that lie beyond the reach of our telescopes, but somewhere on the slippery slope between that and the idea that there is an infinite number of universes, credibility reaches a limit. As one slips down that slope, more and more must be accepted on faith, and less and less is open to scientific verification. Extreme multiverse explanations are therefore reminiscent of theological discussions. Indeed, invoking an infinity of unseen universes to explain the unusual features of the one we do see is just as ad hoc as invoking an unseen Creator..... The multiverse theory may be dressed up in scientific language, but in essence it requires the same leap of faith.....
George Ellis a

George Francis Rayner Ellis, FRS, Hon. FRSSAf, is the emeritus distinguished professor of complex systems in the Department of Mathematics and Applied Mathematics at the University of Cape Town in South Africa. He co-authored The Large Scale Structure of Space-Time with University of Cambridge with physicist Stephen Hawking, published in 1973, and is considered one of the world's leading theorists in cosmology. From 1989 to 1992 he served as president of the International Society on General Relativity and Gravitation. He is a past president of the International Society for Science and Religion. He is an A-rated researcher with the NRF....

George Ellis, writing in August 2011, provided a criticism of the multiverse, and pointed out that it is not a traditional scientific theory. He accepts that the multiverse is thought to exist far beyond the cosmological horizon. He emphasized that it is theorized to be so far away that it is unlikely any evidence will ever be found. Ellis also explained that some theorists do not believe the lack of empirical testability falsifiability is a major concern, but he is opposed to that line of thinking:

Many physicists who talk about the multiverse, especially advocates of the string landscape, do not care much about parallel universes per se. For them, objections to the multiverse as a concept are unimportant. Their theories live or die based on internal consistency and, one hopes, eventual laboratory testing......

Ellis says that scientists have proposed the idea of the multiverse as a way of explaining the nature of existence..... He points out that it ultimately leaves those questions unresolved because it is a metaphysical issue that cannot be resolved by empirical science.....He argues that observational testing is at the core of science and should not be abandoned...

As skeptical as I am, I think the contemplation of the multiverse is an excellent opportunity to reflect on the nature of science and on the ultimate nature of existence: why we are here.... In looking at this concept, we need an open mind, though not too open. It is a delicate path to tread...... Parallel universes may or may not exist; the case is unproved..... We are going to have to live with that uncertainty..... Nothing is wrong with scientifically based philosophical speculation, which is what multiverse proposals are..... But we should name it for what it is....

— George Ellis, Scientific American, "Does the Multiverse Really Exist?"[1]

CLASSFICATION- Cosmologist Mark Tegmark has provided a taxonomy of universes beyond the familiar observable universe....The four levels of Tegmark's classification are arranged such that subsequent levels can be understood to encompass and expand upon previous levels....They are briefly described below...

Level I: An extension of our universe

A prediction of cosmic inflation is the existence of an infinite ergodic universe, which, being infinite, must contain Hubble volumes realizing all initial conditions[2].......

Accordingly, an infinite universe will contain an infinite number of Hubble volumes, all having the same physical laws and physical constants..... In regard to configurations such as the distribution of matter, almost all will differ from our Hubble volume..... However, because there are infinitely many, far beyond the cosmological horizon, there will eventually be Hubble volumes with similar, and even identical, configurations.... Tegmark estimates that an identical volume to ours should be about $10^{10^{115}}$ meters away from us.....

Given infinite space, there would, in fact, be an infinite number of Hubble volumes identical to ours in the universe.....This follows directly from the cosmological principle, wherein it is assumed that our Hubble volume is not special or unique.....

Level II: Universes with different physical constants-

In the eternal inflation theory, which is a variant of the cosmic inflation theory, the multiverse or space as a whole is stretching and will continue doing so forever, but some regions of space stop stretching and form distinct bubbles (like gas pockets in a loaf of rising bread). Such bubbles are embryonic level I multiverses.....

Different bubbles may experience different spontaneous symmetry breaking, which results in different properties, such as different physical constants.....[3]

Level II also includes John Archibald Wheeler's oscillatory universe theory and Lee Smolin's fecund universes theory......

Level III: Many-worlds interpretation of quantum mechanics

Hugh Everett III's many-worlds interpretation (MWI) is one of several mainstream interpretations of quantum mechanics....

In brief, one aspect of quantum mechanics is that certain observations cannot be predicted absolutely. Instead, there is a range of possible observations, each with a different probability. According to the MWI, each of these possible observations corresponds to a different universe. Suppose a six-sided dice is thrown and that the result of the throw corresponds to a quantum mechanics observable. All six possible ways the dice can fall correspond to six different universes.

Tegmark argues that a Level III multiverse does not contain more possibilities in the Hubble volume than a Level I or Level II multiverse....In effect, all the different "worlds" created by "splits" in a Level III multiverse with the same physical constants can be found in some Hubble volume in a Level I multiverse.....
Similarly, all Level II bubble universes with different physical constants can, in effect, be found as "worlds" created by "splits" at the moment of spontaneous symmetry breaking in a Level III multiverse...
According to Yasunori Nomura,Raphael Bousso, and Leonard

Susskind, this is because global spacetime appearing in the (eternally) inflating multiverse is a redundant concept. This implies that the multiverses of Levels I, II, and III are, in fact, the same thing. This hypothesis is referred to as "Multiverse = Quantum Many Worlds"......
According to Yasunori Nomura, this quantum multiverse is static, and time is a simple illusion.....

Level IV: Ultimate ensemble

The ultimate mathematical universe hypothesis is Tegmark's own hypothesis.

This level considers all universes to be equally real which can be described by different mathematical structures.

Tegmark writes:

Abstract mathematics is so general that any Theory Of Everything (TOE) which is definable in purely formal terms (independent of vague human terminology) is also a mathematical structure. For instance, a TOE involving a set of different types of entities (denoted by words, say) and relations between them (denoted by additional words) is nothing but what mathematicians call a set-theoretical model, and one can generally find a formal system that it is a model of.

He argues that this "implies that any conceivable parallel universe theory can be described at Level IV" and "subsumes all other ensembles, therefore brings closure to the hierarchy of multiverses, and there cannot be, say, a Level V....

Brian Greene's nine types

The American theoretical physicist and string theorist Brian Greene discussed nine types of multiverses:[

Quilted

The quilted multiverse works only in an infinite universe. With an infinite amount of space, every possible event will occur an infinite number of times. However, the speed of light prevents us from being aware of these other identical areas.

Inflationary

The inflationary multiverse is composed of various pockets in which inflation fields collapse and form new universes.

Brane

The brane multiverse version postulates that our entire universe exists on a membrane (brane) which floats in a higher dimension or "bulk". In this bulk, there are other membranes with their own universes. These universes can interact with one another, and when they collide, the violence and energy produced is more than enough to give rise to a big bang. The branes float or drift near each other in the bulk, and every few trillion years, attracted by gravity or some other force we do not understand, collide and bang into each other. This repeated contact gives rise to multiple or "cyclic" big bangs. This particular hypothesis falls under the string theory umbrella as it requires extra spatial dimensions.

Cyclic

The cyclic multiverse has multiple branes that have collided, causing Big Bangs. The universes bounce back and pass through time until they are pulled back together and again collide, destroying the old contents and creating them anew.

Landscape

The landscape multiverse relies on string theory's Calabi–Yau spaces. Quantum fluctuations drop the shapes to a lower energy level, creating a pocket with a set of laws different from that of the surrounding space.......

Quantum

The quantum multiverse creates a new universe when a diversion in events occurs, as in the many-worlds interpretation of quantum mechanics......

Holographic

The holographic multiverse is derived from the theory that the surface area of a space can encode the contents of the volume of the region.....

Simulated

The simulated multiverse exists on complex computer systems that simulate entire universes....

Ultimate

The ultimate multiverse contains every mathematically possible universe under different laws of physics......

Contents

REFRENCES

[1] "Does the Multiverse Really Exist?". Scientific American. Vol. 305 no. 2. New York City: Nature Publishing Group..

[2] Tegmark, Max (2003)...."Parallel Universes". In "Science and Ultimate Reality: From Quantum to Cosmos..

[3]Parallel universes. Not just a staple of science fiction, other universes are a direct implication of cosmological observations.", Tegmark M., Sci Am. 2003 May;